WONDER WHALES®

"THE ADVENTURE BEGINS"

Created and Written by **Judith Ellis**
Story Illustrated by **Loren Chantland**
Learn and Play Pages Illustrated by **Rosita Finn**

For permission information please contact:
  Marketing Department
  Wonder Whales, Inc.
  401 N. 3rd Street, Suite 676
  Minneapolis, Minnesota 55401

Wonder Whales is a trademark of Wonder Whales, Inc.

Library of Congress
  Catalog Card Number 95-61055

ISBN 1-887527-25-7

Printed in the United States of America
  First printing June 1995

Created and written by ~ Judith Ellis.

Illustrated by ~ Loren Chantland with airbrush and magic.

Learn & Play pages illustrated by ~ Rosita Finn with marker pens and various media.

Book design by ~ Judith Ellis and Geoff Bush.

Mac prepress production by ~ Mike Zenisek, Peggy Sue Chapman, Linda Muehlbauer and Jane Klesk at The Machanics, Minneapolis, MN.

Color separations by ~ Steve McFarlan and crew at Prepress Management Group, Eden Prarie, MN.

Printed and bound by ~ Berryville Graphics, Berryville, VA.

For my parents, Francis Jurisch and Ingeborg Christine Bush, and my brother Geoffrey Ellis Bush ~ the adventure began with your belief in me.

Judith Ellis

Thanks to all my teachers...

Loren Chantland

THE ADVENTURE BEGINS

O·O IT'S OKAY TO BE ORIGINAL

Mysti Make Your Mark

Cero Save Our Seas

It's A Wonder! The Wonder Whales' names come from their scientific names! O-O, the killer whale, has a scientific name of Orcinus orca. Mysti, the humpback whale, is a Mysticeti whale. Cero, the narwhal, has a scientific name of Monodon monoceros.

It's A Wonder! Delphi, the beluga whale, has a scientific name of Delphinapterus leucas.

The sea is a soup filled with amazing creatures and mysterious life forms...electrifying eels, fantastic fish, and sinister sharks. Yet none are so awesome as that curious quartet of cetaceans* ~ those defenders of the deep ~ known as the Wonder Whales! Don't they make you wonder?

Wonder Word – Cetacean (Se-TAY-shun) Order of marine mammals that includes whales, dolphins, and porpoises.

The waves were up and the waves wouldn't wait. The Wonder Whales were late, but they knew well that the waves wouldn't wait. Mysti, the acrobatic humpback, sliced through the sparkling surf with her far-reaching fins. Cero (SEAR-oh), the shy narwhal, stirred the surf with his twisty tusk.*

It's a Wonder! There really are whales with long, twisty tusks. Some people call them "unicorns of the sea." The tusk is a tooth that grows through the narwhal's upper lip.

Delphi (DELL-fee), the bossy beluga, bounced upon the crests. O-O, the mischievous killer whale, caught curl after curl.

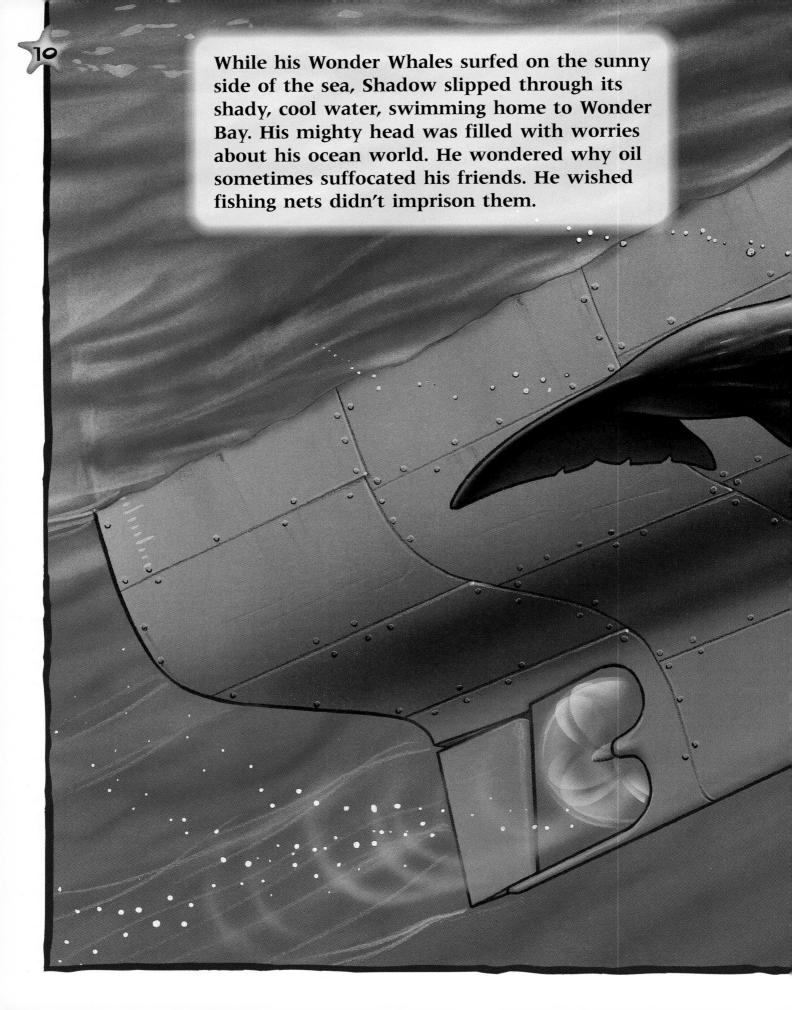

While his Wonder Whales surfed on the sunny side of the sea, Shadow slipped through its shady, cool water, swimming home to Wonder Bay. His mighty head was filled with worries about his ocean world. He wondered why oil sometimes suffocated his friends. He wished fishing nets didn't imprison them.

He wanted to understand why some of the
creatures that live on the land want to watch
whales...and why some want to hunt them.
Shadow was deep in wonder and did not
see the pirate whaleship, the *Dark Voyager*,
watching and waiting above.

Without warning, a single harpoon was fired from the *Dark Voyager*. A shattering explosion was heard. The sea began to shake. And Shadow was lost.

Mysti's sensitive skin felt the sea shiver and she spy-hopped*
on the surface to take a look around.

"What was THAT? Did you feel the sea shaking?" Mysti asked.

"I HEARD it," Delphi answered.

"Feel. Hear. One is like the other to me," Mysti said.

"Forget about hearing and feeling. Look at the sun and think
about getting home. We're late!" O-O called to the others.

Delphi called back, "And we're in trouble! Uh-oh, O-O! Cero!
Mysti! Out of here! Now!"

Wonder Word – Spy-hop: Sometimes whales lift their heads straight out of the water and take a look around.
This behavior is called "spy-hopping."

Home at Wonder Bay, True Blue told her Wonder Whales,
"I was very worried about you, and later we'll talk about why
you were late. But now I'm worried about Shadow. He was
supposed to be home by now too."

Chelo (CHELL-oh), the wise sea turtle, listened.

Mysti whispered a warning to the others, "True Blue looks too
blue. Something must be wrong. Shadow might be in danger.
Maybe it has something to do with that sound I felt ~ or that
feeling I heard."

Delphi sent a signal to O-O, Cero, and Mysti. "Click-chirp-click-
chirp-whistle." The signal said, "Rendezvous at Rhino Reef."

The Wonder Whales swam to Rhino Reef. They felt protected in this secret place. Here they solved problems. Here they found and felt their power.

O-O told the others, "I'm never afraid of anything. I wonder why I feel so afraid for Shadow right now."

"Look O-O! You are either not afraid of anything or you are. Which is it? Choose one or the other," Delphi demanded.

"SHEESH, Delphi, you are so bossy," O-O answered. "Does it have to be one or the other? Right or wrong? Afraid or not afraid? Aren't YOU being kind of black and white?"

Chelo reminded them, "Black and white is not important. Gray is important. Remember why you are here. Think of Shadow."

Mysti said, "But Chelo, we don't know what happened to Shadow. We don't even know where it happened. How can we help him if we can't find him?"

Chelo said quietly, "You will find him. You are Wonder Whales. Remember to listen for the echoes. Remember that what you send out is returned to you."

The Wonder Whales used their powers of echolocation*
and found Shadow hiding in the hull of an old sunken
ship. Wounded and weak, he whispered his story to them.

"It was that ship, the *Dark Voyager*, the one near Skeleton
Cove. They're pirate whalers, and they'll harm you if they
know that I survived. I have their list ~ a list of whales
they have planned for extinction."

Wonder Word – Echolocation (eko-low-KAY-shun) Whales talk to each other and find food by sending out a sound like a click and waiting for its echo to return. This is called "echolocation."

"Extinction*? Wow! No wonder they call it the Endangered Species List*," Mysti said to Shadow.

"Yes, endangered species need protection, not only from pirate whalers, but from pollution too. And I need time to rest and heal. So leave, and remember that no one must know that I'm still alive. When I am well, I will return and the killing will stop," Shadow said before he drifted into sleep.

It's A Wonder! Some whales are so rare that they are on a special government list of animals that need protection so they don't become extinct. **Wonder Word – Extinction** (ex-TINK-shun) No longer existing. If whales become "extinct," there will be no more whales in the world.

The Wonder Whales swam through the seas, warning other whales to watch and be aware. When they gathered at Rhino Reef a few days later, Delphi told the others, "I went back to check on Shadow, and his wounds are worse than he lets on. It will be months before he'll be able to swim on his own."

O-O said, "How many endangered species will be blown into extinction by then? We can't wait ~ doing nothing! We're Wonder Whales! But what? What should we do?"

Shy Cero spoke, "The *Dark Voyager* has the weapons that kill. We need to find a way to disable it so whales are safe again. Let's go to that diving competition at the swimming hole~take a break, clear our heads, take some time to think about this."

"Cero is right," O-O said. He felt powerful as he called out, "Cetacean strength, now and forever! Wonder Whales, GO!"

But before they swam off, Chelo warned O-O, "You O-O, sometimes do not pay enough attention to what is happening around you. Keep your eyes open. Be very careful. You might be a marked whale."

In their excitement, the Wonder Whales hadn't seen Sledge and Zyg circling above them. But the sharks had seen them. They were listening, and they learned that Shadow lived.

"So~Shadow is still alive? Sounds pretty bad off too. Let's find him and finish him off. The guys up there would like that. Maybe they'll throw us some scraps of food." Sledge said.

Zyg saw the cold look of revenge in Sledge's eyes, and felt worried. "Listen, Sledge," Zyg said, "let's go down to Car Wreck Reef. Do a little body work. Bang on some bumpers. Chomp on some hubcaps. We'll figure something out, okay? I don't feel very good about taking on Shadow~those four kids and all."

Sledge snarled at Zyg as they swam to Car Wreck Reef, "What's the matter? Are you a sissy shark?"

"Look, Sledge, I've been thinking," Zyg said, "it's a big ocean. We could search forever and never find Shadow. But those Wonder Whales know exactly where he is. So here's my idea. We watch 'em. We wait. When one of 'em separates from the group, we grab 'em. That will bring Shadow out of the shadows for sure."

Sledge shook his hammerhead, rammed it against a car door, and said, "Yeah, yeah. I still like my idea better. But, okay. We'll do it your way. I guess it's something different."

The Wonder Whales stayed close together as they swam to the diving competition at their favorite swimming hole. Sledge and Zyg watched from afar and waited patiently for one of the whales to be off guard. Opportunity struck for the sharks when O-O's sunglasses fell off and drifted gently to the ocean floor.

O-O told the others, "You go ahead. I'm going to get my sunglasses. I'm so fast, I'll catch up and still get to the dive meet before you do!"

Mysti, Cero, and Delphi sped away as O-O dove deep. Without a whisper of sound, the swift sharks surrounded O-O in a furious frenzy.

Slashing their tails, Sledge and Zyg kept O-O in a spinning prison as they swam toward the *Dark Voyager*. They locked him in a heavy cage and watched as he moved up and down, rising to the surface to take in a breath of air.* His powerful muscles couldn't help O-O now and he felt frustration and icy fear.

It's A Wonder! Whales are like people and need air to survive. They come up to the surface to breathe. Some whales can stay underwater for as long as two hours.

O-O pleaded with Sledge, "Release me. This isn't the answer. Set me free ~ now ~ before it's too late!"

"Too late for what, you big checkerboard!" Sledge rammed his head on the cage.

"Too late to make a difference," O-O said.

Sledge rammed his head on the cage again and spoke coldly. "You don't get it, do you kid? You're a real killer, that's for sure. See I don't CARE! I really don't CARE!"

Sledge swam to Zyg, and the sharks circled below the cage before swimming away.

It's A Wonder! Sharks have rows and rows of teeth. When one falls out, a new one moves up to take its place. Over a lifetime, a shark can use up thousands of sets of teeth.

O-O was left suspended and alone. Again and again he slowly rose for air and lowered himself back below the surface. The rhythm of rising and returning reminded him of the secret Wonder Whales signal, and he began to click-chirp-click-chirp-whistle to himself.

He thought, "I should have listened to Chelo. Kept my eyes open. Click-chirp-click-chirp-whistle. Wonder Whales, Wonder Shmales. Cetacean strength, now and for PHOOEY. Where are you when I need you? Click-chirp-click-chirp..."

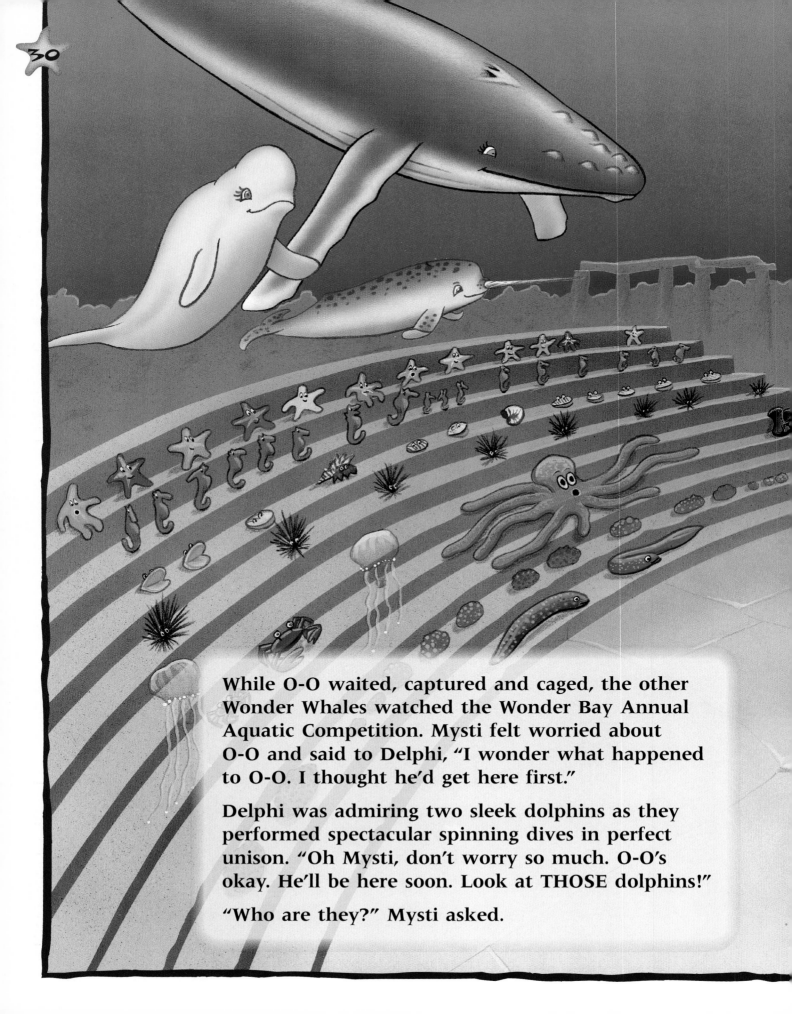

While O-O waited, captured and caged, the other Wonder Whales watched the Wonder Bay Annual Aquatic Competition. Mysti felt worried about O-O and said to Delphi, "I wonder what happened to O-O. I thought he'd get here first."

Delphi was admiring two sleek dolphins as they performed spectacular spinning dives in perfect unison. "Oh Mysti, don't worry so much. O-O's okay. He'll be here soon. Look at THOSE dolphins!"

"Who are they?" Mysti asked.

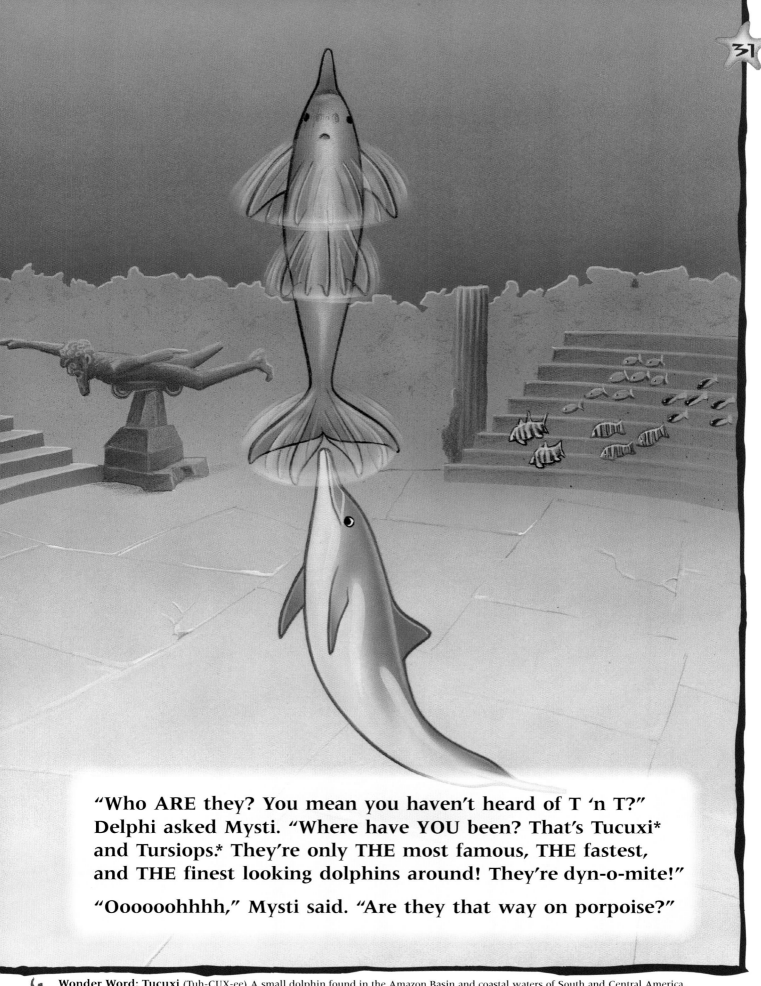

"Who ARE they? You mean you haven't heard of T 'n T?"
Delphi asked Mysti. "Where have YOU been? That's Tucuxi*
and Tursiops.* They're only THE most famous, THE fastest,
and THE finest looking dolphins around! They're dyn-o-mite!"

"Ooooooohhhh," Mysti said. "Are they that way on porpoise?"

Wonder Word: **Tucuxi** (Tuh-CUX-ee) A small dolphin found in the Amazon Basin and coastal waters of South and Central America.
Wonder Word: **Tursiops** (TUR-see-ops) Also called the "bottlenose" dolphin and often seen performing in aquariums.

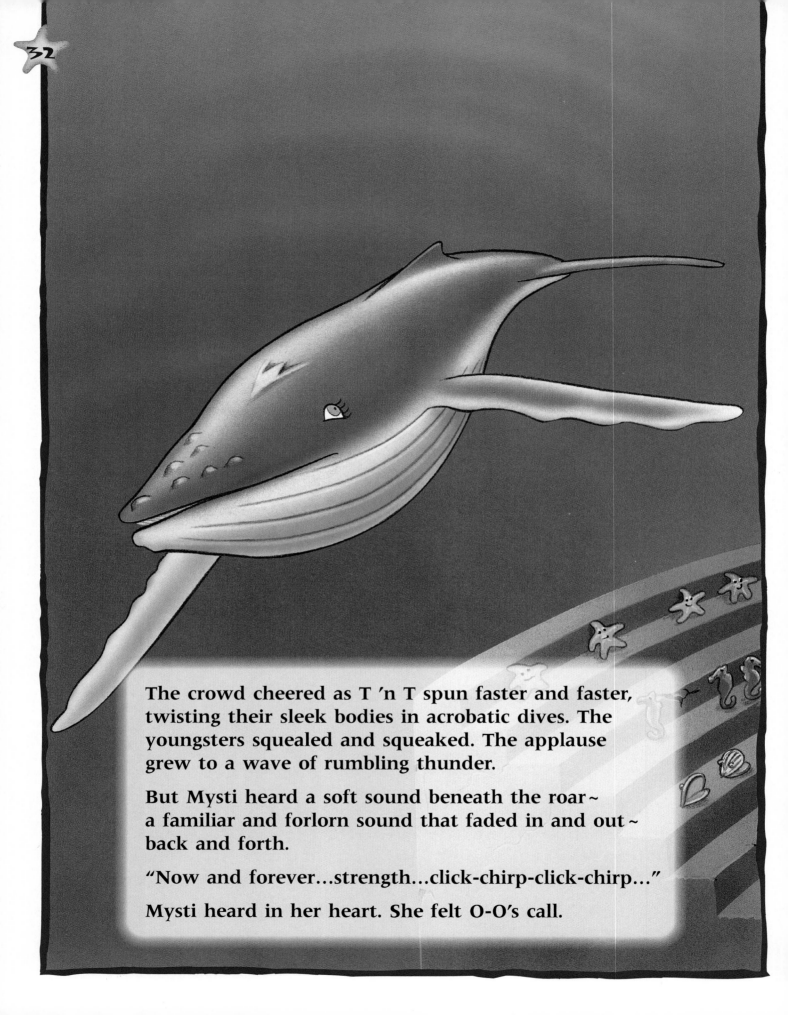

The crowd cheered as T 'n T spun faster and faster, twisting their sleek bodies in acrobatic dives. The youngsters squealed and squeaked. The applause grew to a wave of rumbling thunder.

But Mysti heard a soft sound beneath the roar ~ a familiar and forlorn sound that faded in and out ~ back and forth.

"Now and forever...strength...click-chirp-click-chirp..."

Mysti heard in her heart. She felt O-O's call.

"I hear O-O. He must be in trouble. Let's go! I can follow his signal," Mysti called to Cero and Delphi.

Cero looked at T 'n T and told them, "You come too."

Delphi whispered to Cero, "Not that I mind, but why should THEY come? They aren't Wonder Whales."

"Maybe they are, Delphi," Cero answered. "Maybe they just haven't discovered it yet."

Delphi cried out, "Cetacean strength, now and forever. Wonder Whales, GO!" The Wonder Whales and T 'n T swam off to find O-O.

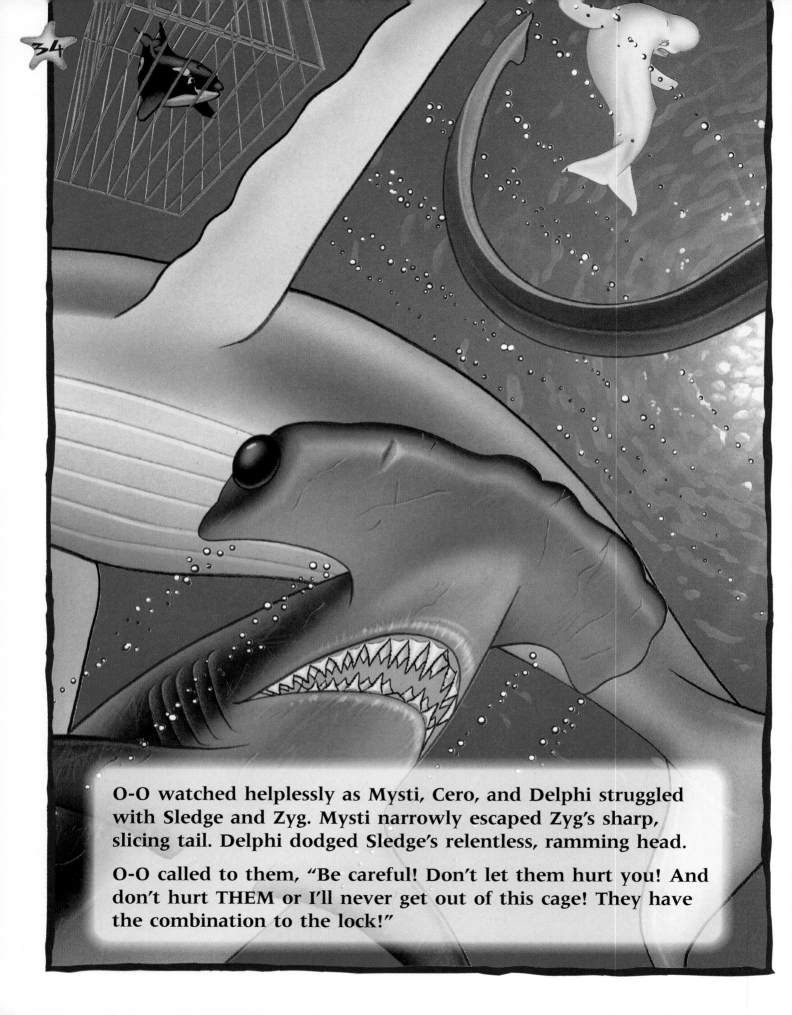

O-O watched helplessly as Mysti, Cero, and Delphi struggled with Sledge and Zyg. Mysti narrowly escaped Zyg's sharp, slicing tail. Delphi dodged Sledge's relentless, ramming head.

O-O called to them, "Be careful! Don't let them hurt you! And don't hurt THEM or I'll never get out of this cage! They have the combination to the lock!"

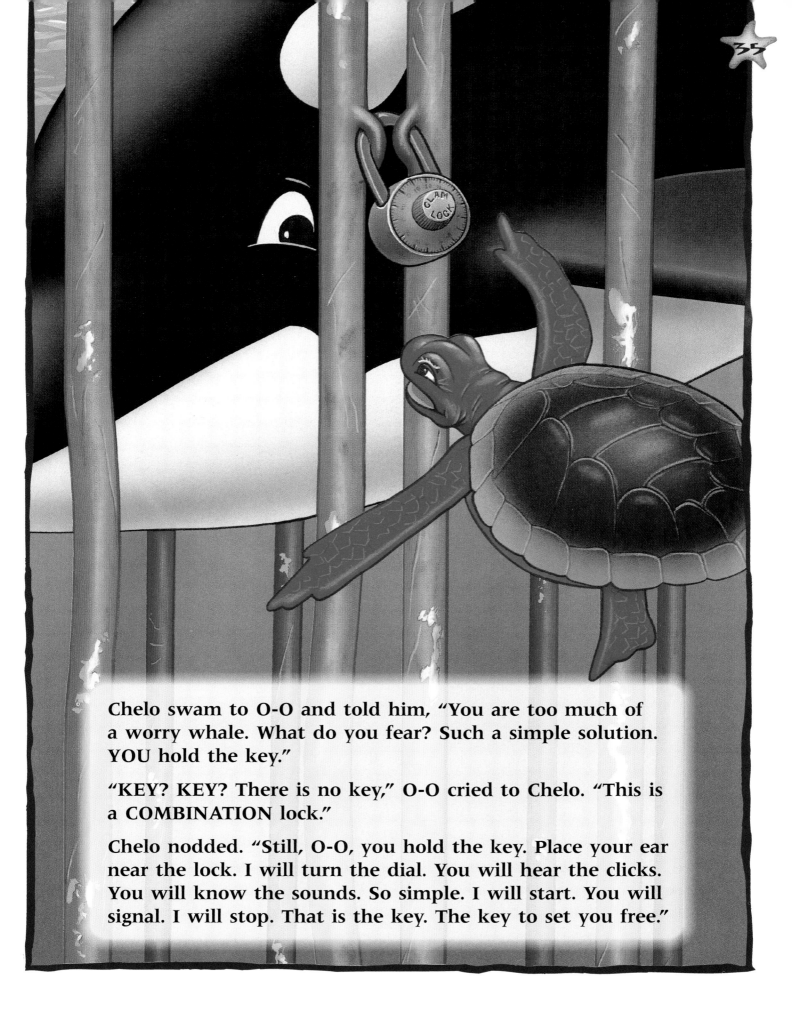

Chelo swam to O-O and told him, "You are too much of a worry whale. What do you fear? Such a simple solution. YOU hold the key."

"KEY? KEY? There is no key," O-O cried to Chelo. "This is a COMBINATION lock."

Chelo nodded. "Still, O-O, you hold the key. Place your ear near the lock. I will turn the dial. You will hear the clicks. You will know the sounds. So simple. I will start. You will signal. I will stop. That is the key. The key to set you free."

O-O swam free and the Wonder Whales listened to Cero as he told them, "If we all work together, we can push the *Dark Voyager* onto that sandbar. Then it can't go to sea to hunt whales."

T 'n T asked, "Why don't we just sink it?"

"This way," Cero answered, "we disable the ship. If there are any people on it, they won't drown. And, we have to think about the ocean. If we sink the ship, it will rust and all the toxic fluids will leak into the sea. Not a good thing for those of us who live here."

The Wonder Whales concentrated and pushed as hard as they could against the hull of the ship. They heard it creak and moan as it settled onto the soft sandbar above them.

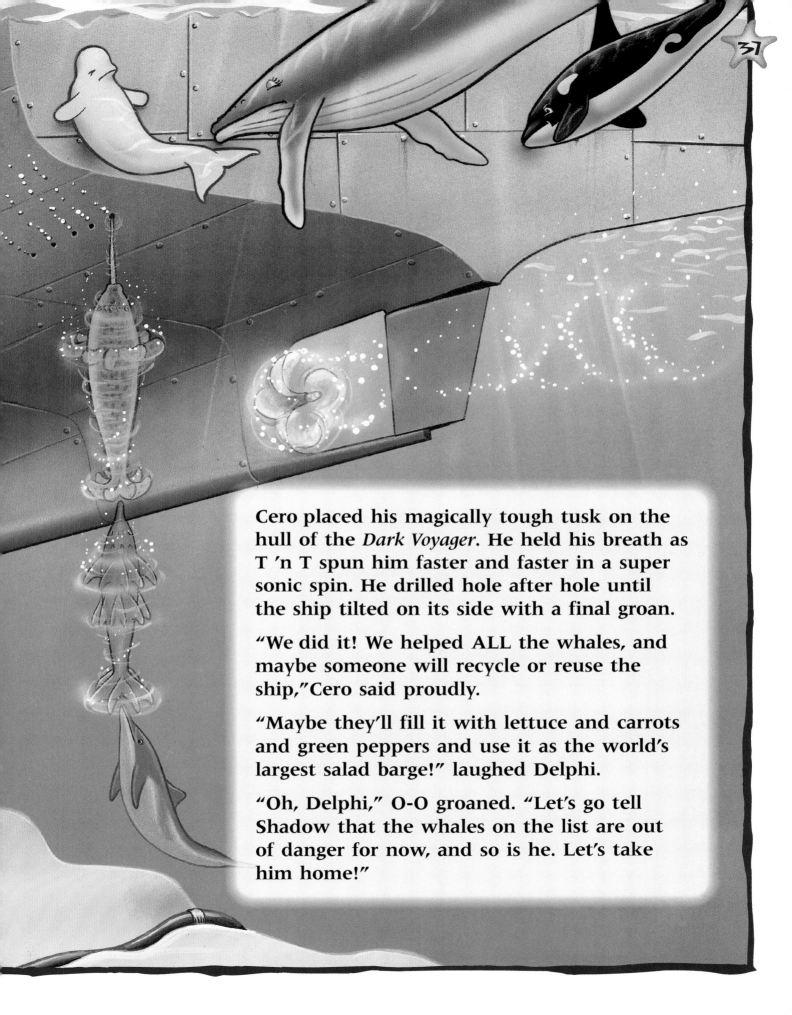

Cero placed his magically tough tusk on the hull of the *Dark Voyager*. He held his breath as T 'n T spun him faster and faster in a super sonic spin. He drilled hole after hole until the ship tilted on its side with a final groan.

"We did it! We helped ALL the whales, and maybe someone will recycle or reuse the ship," Cero said proudly.

"Maybe they'll fill it with lettuce and carrots and green peppers and use it as the world's largest salad barge!" laughed Delphi.

"Oh, Delphi," O-O groaned. "Let's go tell Shadow that the whales on the list are out of danger for now, and so is he. Let's take him home!"

"Wow, Shadow," Delphi chirped noisily. "You wouldn't BELIEVE what happened to us. It was unreal! There were these sharks and they put O-O in a teensy, weensy cage, and we all worked together to rescue him. And we pushed the *Dark Voyager* up onto a sandbar!"

"What a day! I think I need braces," Cero said as he rubbed his twisty tusk.

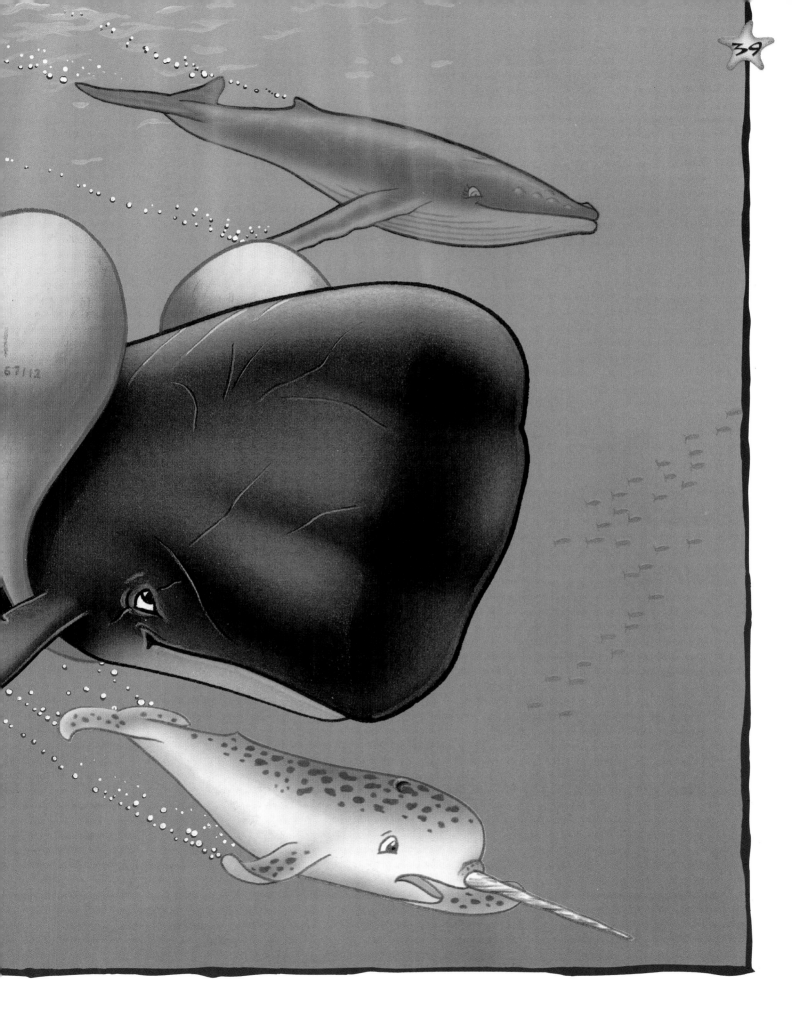

"I'm just happy that all of you are safe," True Blue told Shadow and the Wonder Whales.

"Me too," O-O said. "Hey, did anyone else notice that weird cable? It looked like a big, black snake slithering under the sand. I wanted to check it out, but there were too many distractions."

"I saw it too," Cero and Delphi chirped together.

"I heard it...umm...felt it. It's sending out hums and clicks and...," Mysti began.

Chelo interrupted, "Yes, it sends out hums and clicks. And remember that what you send out is returned. Perhaps it is nothing but a sea snake that is miles and miles long. Perhaps it is something else. It is what it is. But what is it that it is? Doesn't it make you wonder?"

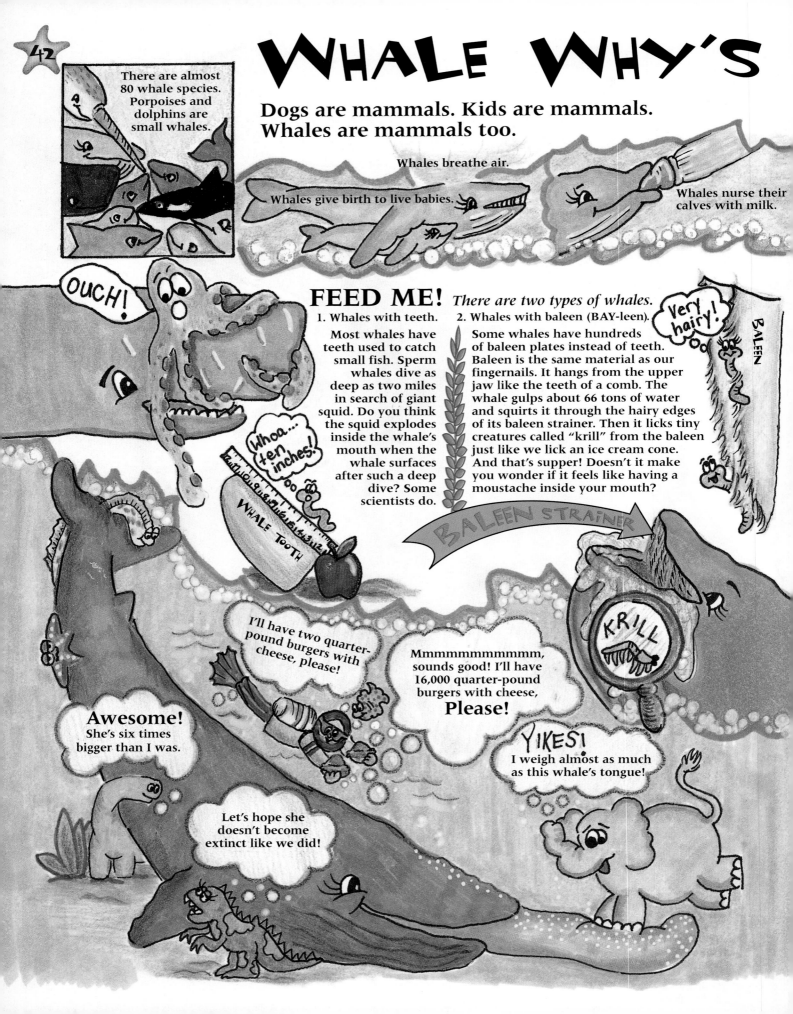

WHALE WATCH

**Do you WONDER where the WHALES are?
Whales are everywhere in the world!**

You can find whales at the top
of the world, at the bottom, and
everywhere in between! How many
whales can you find in this picture?
How many whales can you find that
belong back in the water?

S.S. CERO
S.O.S.

Answers: There are 18 whales in this picture. 12 whales belong back in the water.

AMAZING WHALES

The oceans are filled with eerie squeaks, clicks, and cries. Humpback whales sing haunting songs, and all whales send messages to each other by slapping their flippers and flukes* on the surface of the sea. Whales also use echolocation (eko-low-KAY-shun) to "talk" to each other. They send out a sound and listen for its echo to return. This also helps them find food in the dark depths of the sea. Sperm whales, like the one at the beginning of this maze, use echolocation to find their favorite food, the giant squid. They must dive almost two miles to enjoy a tasty lunch!

Can you help the whale echolocate for its lunch?

Use ballpoint pen.

START HERE TO FIND LUNCH!

Wonder Word – Flukes (FLOOKS) Whale's tail.

STRANDED!

Do you wonder why whales strand themselves on shore? So do scientists! They wonder if the whales are sick or if groups of whales follow a leader onto the beach. Everyone wonders WHY, even after people take them out to deep water, hundreds of whales will swim back to land again and again. Count the whales, count the people, and fill in your answers to read the poem "SAFE AND FREE."

It's A Wonder!

Whales breathe air but they can't survive on land for very long. They are so heavy that they are crushed by their own weight when they are out of the water!

"SAFE AND FREE"

____ whales are stranded on shore!

____ whales are still in the sea!

____ people are trying to help all

____ whales swim safe and free!

SAVE OUR SEAS PROJECT!

Ask a grown up to help you use a scissors safely.

Every year our friends in the ocean accidentally eat or become tangled in the plastic rings from our soft drink packages. You can help save all sea creatures and the earth by cutting the rings before throwing them away in the garbage.

WE ARE THE WONDER WHALES
(OFFICIAL THEME SONG)

We are the Won-der Whales. Hear our waves of
thun-der. Learn of our fan-tas-tic tales!

Don't we make you won-der?
I'm the long-winged
My friends call me
I'm the bos-sy
I'm the twisty-tusk

Mys - ti.
O - O
Del - phi
Ce - ro.
These words are my spark.
and tell me I'm not ty-pi-cal.
and I have the con-fi-dence
Trust me on this please.

E - ven when it's risk - y,
I just want the world to - know.
to never let a chance go - by
Ev - ery - one's a he - ro
go out and make your
It's o-kay to be
to dare to make a
when they help to save our

mark!
o - rig - in - al!
dif - fer - ence!
seas!
We are the Won-der Whales.

We are the Won-der Whales. We are the

Won-der Whales.

Music by ~ Jason Bush   Lyrics by ~ Judith Ellis
© 1995 Jaybird Songs, ASCAP

WHALE NOTES TOO!

TAKE NOTES! WRITE A SONG! CREATE A POEM! TELL A STORY ABOUT WHALES!

Thanks for reading Wonder Whales Book 1.

See you in Book 2, O-O Beneath The Arctic Ice.